CORPSE FLOWER

ASHLEY GISH

look!

CORPSE FLOWER

CREATIVE EDUCATION • CREATIVE PAPERBACKS

Published by Creative Education
and Creative Paperbacks
P.O. Box 227, Mankato, Minnesota 56002
Creative Education and Creative Paperbacks are imprints of
The Creative Company
www.thecreativecompany.us

Design and art direction by Blue Design
Edited by Kremena Spengler

Photographs by Getty Images/Adi Prima/Anadolu, 9, Afriandi, 2, 15, Fadil
Aziz, 1, 21, Mindy Schauer/Digital First Media/Orange County Register, 11;
Shutterstock/Tanveer Anjum Towsif, cover, 23; Wikimedia Commons/AbZahri
AbAzizis, 16, A.J.E. Terzi, 15, Edrinaldi, 5, Fbianh, 6–7, KugelaP, 13, 14–15,
Neisyaranifauzia, 20, Ragesoss, 22, RahmadHimawan Photography, 18–19,
Rhododendrites, 10, Richard J. Rehman, 8, 17, 23
Every effort has been made to contact copyright holders for material
reproduced in this book. Any omissions will be rectified in subsequent
printings if notice is given to the publisher.

Library of Congress Cataloging-in-Publication Data
Names: Gish, Ashley, author.
Title: Corpse flower / Ashley Gish.
Description: Mankato, Minnesota : Creative Education and Creative
   Paperbacks, [2026] | Series: Look! | Includes bibliographical references
   and index. | Audience: Ages 6–9 | Audience: Grades 2–3 | Summary: "A
   giant exotic plant that smells like dead flesh? Elementary-level readers
   will learn all about the corpse flower with this striking introduction
   to the unusual plant's life cycle"– Provided by publisher.
Identifiers: LCCN 2024044285 (print) | LCCN 2024044286 (ebook) | ISBN
   9798889895855 (library binding) | ISBN 9781682777510 (paperback) | ISBN
   9798889896654 (ebook)
Subjects: LCSH: Amorphophallus titanum–Juvenile literature. |
   Amorphophallus titanum–Life cycles–Juvenile literature.
Classification: LCC QK495.A685 G54 2026 (print) | LCC QK495.A685
(ebook)
   | DDC 584/.442–dc23/eng/20250109
LC record available at https://lccn.loc.gov/2024044285
LC ebook record available at https://lccn.loc.gov/2024044286

Printed in India

TABLE OF CONTENTS

No other plant in the world is like the corpse flower. It can grow to be more than 10 feet (3 meters) tall. It blooms for just a few days every two to ten years. When it blooms, it smells like rotten meat.

Corpse flowers grow well in warm, damp places. Experts grow them in **botanical** gardens. Many people go there to see these strange plants.

botanical: having to do with plants

9

10 The corpse flower corm can weigh up to 250 pounds (113 kilograms).

A corpse flower starts as a seed. After three to five months, a **corm** grows out from the seed. One long leaf grows from the corm. After four weeks, the leaf may be 15 feet (4.6 m) tall.

After 12 to 18 months, the leaf dies. The corm rests for about six months. Then a new leaf grows. This can happen over and over again for 10 years!

corm: a round underground stem that stores energy for growing plants

Finally, the corm stores enough energy to make the first flower. A different leaf starts growing. It's called a **spathe**. It protects the flower as it grows.

When the spathe opens, a huge spike grows. This spike is called a **spadix**. It heats itself up to about 98 degrees Fahrenheit (36.6 degrees Celsius). That's really hot! It smells terrible!

spathe: a special leaf that grows around a flower to protect it

spadix: a stem with hundreds of tiny flowers grouped closely together

13

14

The bad smell attracts flies. This is because flies feed on dead animals. The flies crawl down into the flower, looking for food. **Pollen** sticks to their bodies.

pollen: yellow powder made by plants that is used to make other plants

The flies find no food. They look inside a different corpse flower. Pollen falls off of their bodies onto the next corpse flower. This is **pollination**.

pollination: when insects carry pollen from flower to flower so plants can make fruit and seeds

The bloom on a corpse flower stays open for up to three days. Then it dries up and falls over. The fruit grows. About a year will pass before the fruit is ripe.

Nearly 500 fruits grow on each plant. Each fruit has two seeds inside. Birds called rhinoceros hornbills eat the fruit. But they poop out the seeds. More corpse flowers grow from the seeds.

RHINOCEROS HORNBILL

Corpse flower fruit is
poisonous to humans.
17

Scientists agree that corpse flowers need help. Some people harm them by cutting down their forest homes.

Other people believe that corpse flowers are dangerous. This is because the leaf poking out of the ground looks like a snake's head! These people cut down the corpse flowers.

18

In 2024, fewer than 1,000 corpse flowers were left in the wild.
19

A corpse flower plant can live up to 40 years.

Life Stages
Seed
Corm
Leaf
Dying back
Spathe
Spadix
Fruit

The corpse flower grows from a seed. A corm grows from the seed. The corm sprouts a leaf. The leaf grows and dies many times. After that, the plant grows a flower. The stinky bloom lasts for a few days. Then the plant grows fruit. A bird eats the fruit. It poops out the seeds. New plants grow from the seeds.

21

Corpse flowers in
botanic gardens have
had names like Darth
Vapor and Stink, Stank,
and Stunk.

22

Corpse flowers are rare in the wild. But they grow well in more than 80 botanical gardens around the world. The corpse flowers in botanical gardens are given funny names. One of the tallest corpse flowers was Stankosaurus Rex. It bloomed in 2023 at The Huntington in San Marino, California. It was more than 8 feet (2.4 m) tall!

23

READ MORE

Einstein, Tamara. *Weird Plants*. Tukwila, WA: KidsWorld Books, 2021.

Griffin, Mary. *Fantastic Plants: Stink Corpse Flowers*. New York: Powerkids
Press, 2022.

Kaiser, Brianna. *Weird Plants*. Minneapolis: Lerner Publishing, 2023.